The Structure of True Reality

"The path to a better world can only be found by embracing True Reality - Individual by Individual"

Charles R Neff

Copyright © 2017 by Charles R. Neff

All rights reserved including the right to reproduce any portion of this book in any form.

Self-Published

Cover design by Mary Neff

ISBN-13: 978-1974473472
ISBN-10: 1974473473

To my wife who has shared life's experiences with me o'er these many years.

Contents

Introduction 6

Part 1: Elements of True Reality

Background 9

Reality Definitions.. 12

Identifying True Reality................................... 14

Criteria for Fundamental Elements...................... 17

Elements of Physical Reality

Fundamental Elements..................................... 19

Derivative Elements... 32

Elements of Psychological Reality

Practical definition of psychology....................... 34

Overview of Psychological Reality..................... 36

Psychological Element Constraints..................... 41

Fundamental Elements..................................... 43

Derivative Elements... 50

Elements of Transcending Reality

Background Summary....................................... 58

Transcending Elements..................................60

Part 2: Using True Reality to Understand and Address World and Personal Issues

Purpose of Part 2..70

Examples

The Individual in the Scheme of "Everything"..........72

Inalienable Rights: Individual vs Group..................74

Religion..76

Education...78

The United States Constitution..............................82

Controlling the "Peoples Representatives"...............84

Term Limits..87

Congressional Rules/Procedures...........................89

Congressional Operational transparency..................91

Compensation for Elected Officials........................93

Controlling the Supreme Court.............................95

Concluding Statement.................. 98

Addendum

Elements of True Reality (listing).......................99

Introduction

The psychology of man — psychology defined as the mind/behavior relationship — has changed little over the span of human existence. People act in accordance with their rationalized nature (i.e. their inherited nature modified by acquired beliefs, experiences, knowledge and reasoning) interacting with their perceived reality. This psychological relationship holds true through all stages of man's development (assuming no physical damage to the brain).

An individual's beliefs, experience, knowledge and reasoning can all — to some degree — be influenced by others. Through this influence both the individual's nature and perception of reality can be manipulated. Because one's nature and perceived reality dictate all human actions — and can be manipulated — they have proven to be a rewarding battleground for those seeking to influence the actions of others.

Essentially, the battles for manipulating psychology are waged around the concept of perceived realities. The battle consists of creating "realities" that serve to produce desired actions, with contending parties promoting different "realities". It is not necessary for a "reality" to be true, only that it be perceived as true. Most of the world's problems, both personal and societal, can be traced to false realities that are accepted as truth.

It is a central premise of this book that grounding human psychology on True Reality provides the only sustainable path for improving the human condition. This premise is itself derived from True Reality. It is further premised that only the deductive method of analysis provides an unambiguous means for defining True Reality (within the limits of current knowledge). The deductive method, as applied to reality, consists of first identifying and verifying the individual truths of reality and then assembling these truths into a framework that logically supports True Reality as a whole. Experience convinces that other methods of analysis, as applied to reality, too easily breakdown due to emotions, issues of philosophy, ignorance as to what is true, or simply reasoning too complex or convoluted to permit unambiguous conclusions.

It is the complexity of reality that opens the door to false realities. It is easy to construct false views of reality that appear both plausible and desirable – providing the "truths" on which they rest are not carefully examined. It is the truths of reality that expose false realities: *no view of reality can be true if it conflicts with any individual truth of reality.*

Of course, truth alone is not sufficient to ensure acceptance of anything, reality included. History amply demonstrates that truths that conflict with existing beliefs or interfere with one's desires are not readily accepted. Considering what many people now perceive as reality – such being fostered by long-term indoctrination by false realities – replacing the false with the true requires more than simply stating the truth. It requires True Reality to meet the criteria of being unambiguously based, clearly stated, logically consistent and independently verifiable. Where these criteria are met all people of good intentions will ultimately accept the truth.

The primary objective of this book is to present the core of True Reality in a manner that meets the above criteria. A secondary objective is to present a few examples of applying True Reality dictates when addressing world and personal affairs. In pursuing these objectives I strive to be as concise as, in my judgement, a subject permits. The intent of being concise is both to aid in making objective true/false judgements on the information presented and to permit a relatively uncluttered picture of True Reality's core.

I make no claim of originality in defining the basis of True Reality. All of the information presented in this book is derived from sources in the public domain, available to anyone in a free society. I am simply presenting this information stripped down to the individual truths involved, thereby breaking a complex reality into more logically verifiable pieces. This is standard procedure for understanding any complex subject. Once the individual truths are established then True Reality, as a whole, must conform to their dictates. False realities cannot survive such analysis.

Part 1: Elements of True Reality

Background

Consider this partial list of current world problems: great disparities in wealth and education; religious (sort of) conflicts; ruthless dictators; corrupted politics; scientific discoveries that contradict long established beliefs, organized terrorism, weapons of mass destruction and environmental/health doomsday projections.

Actually, except for scale, these problems are all old problems. They have existed either continuously or intermittently throughout recorded history. For proof of this one needs only to review earth's geological history and the rise and fall of past civilizations/nations. But, old or not, these problems (and more) remain a source of conflict and tensions in today's world.

To the afore listed problems add in the tensions, felt by both individuals and societies, that are a consequence of today's rapidly evolving technologies and knowledge. Adapting to changing technologies and knowledge requires not only learning the new technologies and information but also frequently requires modifying one's behavior and beliefs. Such adaptation does not come easily to everyone. Those not successfully adapting face increasing disadvantages, both economic and social, that can result in strong personal resentments.

Finally, factor in the communication revolution which makes it possible for anyone, organization or individual, to reach large numbers of people with whatever communication they choose, true or not. The result is a constant flood of information conveyed in an undifferentiated mix of the true, the false and the biased. The bulk of this undifferentiated mix inevitably consists of false or biased information as for each truth there can be an unlimited number of false or biased statements addressing the same subject. The one thing this communication revolution carefully does not provide is a ready means for verifying the validity of the information being conveyed.

This combination of problems, tensions and free-for-all communications creates strong emotions and intellectual confusion – a situation made to order for those wishing to deceive and exploit. For the individual the task is to sort through the jumble of information to find the truth. This task is never easy.

Despite the difficulty in obtaining true information the individual who does not base his/her opinions, beliefs, etc. on truth has little hope of making right decisions. Even a lingering uncertainty as to what is true acts to weaken life's anchor points, not only to what is true or false, but also to what is good or bad, moral or immoral and all the other decisions that bring meaning and value to life.

Unfortunately, even knowing the truth does not always guarantee good decisions – man is not always rational nor of good intentions – but only truth removes ignorance from the process.

With the knowledge and technology now available there has never been greater potential for advancing the welfare of the world – or greater potential for harm. **Which it will be depends primarily on the prevailing world psychology interacting with the prevailing worldview of reality to** produce actions – helpful, ineffectual or harmful actions.

Reality Definitions

In a totally objective world a proposed reality would be either true or false. However, in practice realities have to be <u>perceived</u> by the individual as true or false to have any impact on his/her behavior. Unfortunately, such perceptions are seldom totally objective, which results in perceived realities that are influenced by all sorts of inputs - some true some not. Thus, for practical purposes there are three types of realities:

1) **True reality** is a reality that is consistent with all known physical and psychological truths. The individual truths of reality are universally applicable and are unchanging within the limits of man's present knowledge.

2) A **False reality** is any proposed "reality" that when objectively analyzed does not meet the requirements of a true reality. Such realities have existed throughout history either as a result of ignorance or by design. With the information now freely accessible (at least in the free world) it should be difficult for false realities to exist. Unfortunately, false realities continue to enjoy wide acceptance.

3) A Perceived reality is one that an individual rationalizes as true. For most (if not all) people their perception of reality contains a mix of true and false components. Each individual creates his/her own unique perceived reality, though much may be held in common with others. Perceived realities are not static but change as the individual's knowledge and life experiences accumulate. The importance of the perceived reality derives from the fact that it is this perception, interacting with one's nature, which dictates an individual's behavior/actions.

There are a large number of problems that derive from perceived realities that are false. Attempting to correct each individual problem, even where successful, accomplishes little more than winning skirmishes while losing the war. To achieve lasting success in defeating false realities such "realities" must be attacked all at once and at their roots. Only direct confrontation with true reality provides the means for such an attack.

Identifying True reality

Aside from physical evolutional developments, the history of man is the history of actions taken. Understanding both the prevailing psychology and the perceived realities that produced these actions, and how such psychologies and "realities" came to be, is the key to learning the lessons of history. When looked at from this perspective history clearly shows the damaging influence of ignorance, even with the best of intentions. It also clearly shows how both psychology and "reality" were -- and still are -- being manipulated.

As stated earlier, true reality is too complex to be understood except through deductive reasoning based upon the truths of reality. These truths set both the structure and boundaries for understanding reality. Therefore, the first task for those seeking to understand reality is to identify the truths involved.

In order to avoid the semantic difficulties encountered when evaluating proposed "truths" for validity I will substitute the term "element" for a proposed "truth". An element is then either true or false, avoiding the awkwardness of having true "truths" and false "truths". A True Reality can then be defined as one comprised of only true elements. Such elements, in total, represent the individual facts, beliefs, principles, etc. that form the basis for a true view of reality.

Determining how reality manifests itself is the role that science and experience serve. In pursuing this role man is discovering many aspects of physical reality through science, and many insights into man's psychology through the experience of observing and documenting man's behavior in various situations. Applying this information to man's activities has proved both beneficial and detrimental – depending on how such information is used.

Fundamental Elements are the foundation of True Reality. Such elements are the starting point from which all aspects of the universe – life included – have developed. However, while fundamental elements provide the foundation for reality they do not directly explain the structure of reality as we experience it. To bridge the information gap between what the fundamental elements dictate and how reality actually manifests itself, it is useful to identify additional elements of True Reality. These additional elements can be categorized as Derivative Elements and Transcending Elements.

Derivative Elements are logical consequences of the Fundamental Elements but – as man's logic is often suspect – only Derivative Elements that are verified through the lens of experience can be certified as true. Derivative elements are more limited in scope than Fundamental Elements but more directly relatable to everyday experience. Derivatives of derivatives can also be logically extracted to obtain elements even more closely related to every day experience. However, for the purposes of this book I will limit discussion to first derivatives.

Transcending Elements are the experience and reason based principles that must be satisfied if man is to improve the human condition and the world in which we live.

All elements must be phrased such that unambiguous true/false evaluations are obtainable.

To achieve this end each element is restricted to stating a single fact, belief, principle or concept. This restriction is applied so that the reasoning used in evaluating an element for truth — by virtue of focusing on a single, isolated claim — will also be focused. This focus permits unambiguous true/false judgments as to the truth of the claim.

I propose that you, the reader, subject each and every element I advance to your own analysis to determine your own true/false judgment. If, in your judgment, you find an element to be false then I further propose that you replace it with your own element, along with your arguments as to why your element is true. When all elements are so treated then the framework for your own perceived reality will be revealed. This framework is likely to prove educational both to yourself and to others should you choose to share it.

Criteria for Fundamental Elements

Fundamental elements can be broadly grouped into two basic types: those pertaining to the physical world and those operating in the psychological realm. Physical fundamental elements are based upon scientifically established laws and properties of the universe. These laws and properties are themselves derived from physical measurements of various aspects of the universe. However, psychological elements, by their nature, have no properties that are physically measurable. Instead, psychological elements rest upon observed human behavior, historically and/or scientifically documented, and verified through continued experience.

Psychological perceptions are often dependent upon physical perceptions in ways that are not obvious. Alternately, psychological perceptions can influence how one views the physical world. Because of these interactions it is advantageous to first identify the physical elements before delving into the psychological. By first identifying the scientifically based, objectively proven physical elements (truths) one can with certainty discard psychological perceptions that conflict with these truths. The reverse process, discarding physical elements that conflict with one's subjectively established psychological perceptions, offers no similar certainty of a correct decision.

In what follows a set of true fundamental elements that deal with key aspects of reality will be formulated. Each element will be associated with the reasoning that justifies its inclusion in the true category. The criteria used for determining truth will be:

- The physical elements, to be true, must be founded on uncontested scientific discoveries and/or measurements that are not relative to anything other than the universe in which we dwell.

- The psychological elements, to be true, must be compatible with the physical elements and be universal traits of human behavior proven through experience over the entire course of human history (as much as we know).

Elements of Physical Reality

Fundamental Elements of Physical Reality

Perhaps no subject is more misunderstood by more people than the physical realty of the universe. Part of the reason this situation exists is simply because many of the truths underlying physical reality are relatively recent discoveries established in the past 200 years or so. While 200 years may not seem to qualify as "recent", it is a short time to replace beliefs held for millennia. Not only must individuals reconcile the new scientific truths with their current beliefs but must do so in an environment in which many entities – those having a vested interest in current beliefs – actively resist change. The result is that acceptance of fundamental changes to our perceptions of physical reality tend to occur slowly. So it has been throughout history.

So what is the true nature of the physical universe? Consider the following statements, which are formulated to be true fundamental elements:

1. The origin of the universe is forever <u>physically</u> unexplainable.

2. Everything in the universe consists of energy in one form or another.

3. The fundamental nature of energy is unknown; only <u>some</u> of the forms that energy can take are known.

4. The original source of all energy is unknown.

5. What we perceive as solid matter is almost entirely empty (of matter) space.

6. Empty (of matter) space is filled with energy.

7. The passage of time is not the same everywhere.

8. All life is initiated by, and dependent upon, the consumption of energy.

9. The evolution of all life forms is an established fact but the force(s) driving evolution remain unresolved.

These nine elements were selected to form the core of true physical reality. The next step is to establish the validity of the individual statements so as to justify their status as true fundamental elements of reality.

Element 1: *The origin of the universe is forever physically unexplainable.*

The barrier to explaining the origin of the universe resides in the logical impossibility of explaining how something (the universe or something before) can be created from nothing or, alternately, explaining how something (the universe or something before) has always existed. These two options are what one always arrives at by asking of any proposed starting point of creation: "What existed prior to this point"? Eventually no concepts other than "something came from nothing" or "something always existed" can be conjured to provide an answer. However, neither of these concepts is physically explainable under any logic or scientific rationale. If you doubt these are the only two concepts ultimately available, try questioning, as above, any creation scenarios you can imagine.

Inherently, then, all scenarios claiming to explain creation — scientific, theological, philosophical, mystical, magical, etc. — share a common conclusion. That conclusion is that, somehow, "something" can indeed be created from nothing or that "something" has always existed. This shared conclusion does not require belief in the existence of a God (or gods) as currently conceived, but does require belief in a power indistinguishable from that of a God. **Therefore, at the instant of creation, all scenarios of creation invoke -- either explicitly or implicitly -- a God Power behind the event. There are no other alternatives.**

Once past the instant of creation the role of this God Power in the ongoing "Everything" — including the universe as we experience it — can be logically subject to differing interpretations. **However, creation was the seminal event in which all aspects of the universe, life included, were made possible if not inevitable.** Therefore, the minimum role of this God Power is that of setting the rules for matter and energy interactions and then letting these rules produce what we see. However, a more active role for the God Power, such as guidance or direct intervention, cannot be logically or historically ruled out. **Either way a God Power created "Everything" and now guides, passively or actively, the evolution of that "Everything", including the universe and all life within it.**

Contrary to some beliefs, science possesses no tools, methods or rationale for explaining creation, nor even why the fundamental properties and laws of the universe are what they are. Science can only discover the existing properties and laws of the known universe and attempt to understand how these operate in creating what we perceive as physical reality. Science can then attempt to utilize the various known properties and laws of the universe to achieve desired ends. **To claim or imply that science can do more than this – or ever can do more than this – is a key deception in many false realities.**

Element 2: *Everything in the universe consists of energy in one form or another.*

Until approximately the end of the nineteenth century science held that everything in the universe was either matter or energy and that neither could be created nor destroyed. Subsequent advances in both theory and experiment suggested that matter and energy were related. This thinking led to development of the atomic bomb, which, when detonated, spectacularly demonstrated that matter can indeed translate into energy according to the famous equation $E=mc^2$.

Going in the other direction, turning energy into matter, we have the evidence of subatomic particles being created from energy, either naturally or in particle accelerators, and then either disappearing after giving up some discrete quantity of measurable energy in the form of energy "waves" or combining with other particles to form new particles. Indeed, such particles are only detectable by some consequence of the energy they possess.

The smaller "particles" have interesting observable properties such as the ability to behave as either particle or wave, depending on how they are measured. Since all matter is made up of these subatomic "particles" all matter consists of energy.

As matter is now recognized as a form of energy, the fundamental conservation law of the universe may be expressed as: "Energy is neither created nor destroyed, it only changes form". However, in everyday interactions matter and energy do not perceptively exchange and, as their measurable respective properties (beyond subatomic particles) are so different, it is simply convenient to continue treating matter and energy as separate entities in most everyday calculations.

The foregoing greatly condenses the total body of knowledge verifying the relationship between matter and energy but greater detail is readily available to anyone interested in delving deeper. In any event, science has unequivocally established by theory, experiment and practice the truth of Element 2.

Element 3: *The fundamental nature of energy is unknown; only some of the forms that energy can take are known.*

Another way of stating the conservation of energy law is: The total amount of energy in the universe (which includes energy in the form of matter) remains constant; it only changes form. Common but not the only examples of known energy forms include mechanical, electrical, chemical, heat, light, molecular and nuclear. Countless measurements of the conversion of one form of energy to another have been made and all verify that energy is conserved (remains constant). But no one knows why it is conserved or, in many cases, what the transfer mechanism is or even if conservation remains true under extreme conditions or more accurate measurements.

This last "if", relating to the accuracy of measurements, is more important than many scientists admit. Recent more accurate measurements of the distribution and movement of mass within the universe are not currently explainable without the presence of a large amount of "dark mass" and "dark energy" (dark meaning invisible). The amount of dark energy and dark matter required is larger than the estimated total of detectable matter and energy in the universe. This is not a trivial thing to miss in light of all the measurements and resulting theories of the past. It is highly probable that there are forms of energy all around us of which we are unaware.

There are other observed phenomena, particularly at the quantum level, that point toward the existence of energy forms and interchange mechanisms now unknown. This does not detract from the usefulness of what is known but puts our knowledge in perspective: We know quite a bit, know our knowledge is incomplete, but have no way of knowing how incomplete.

Element 4: *The original source of all energy is unknown.*

The prevailing scientific theory describing the origin of the universe is the "Big Bang". The Big Bang theory proposes that the universe originated from an explosion of energy of unimaginable magnitude. As the energy from the explosion expanded and cooled, all the physical laws governing the universe were set and all matter in the universe coalesced, perhaps still coalescing, from a portion of this energy.

Extrapolations from extensive measurements made on our present day universe tend to support the expected consequences of a "Big Bang" explosion. However, the theory offers no clue as to how this energy originated. This is a rather significant omission considering that the entire scientific rationale depends upon cause and effect, and no cause is included in the Big Bang theory.

All objective scientists recognize that the energy behind the Big Bang is scientifically unexplained. Because they cannot explain the source of this energy most simply ignore it and turn their attentions toward the consequences of this energy. In practice they publish their findings as if the Big Bang created the universe, seldom alluding to their ignorance as to how the energy originated. This practice, inadvertent or deliberate, has created the illusion in the minds of many that the Big Bang is the starting point of creation.

Using rigorous scientific reasoning the best that can be said about the Big Bang, if real, is that it is an effect brought about by some unknown cause. This unknown cause is either a source of energy that is far greater in magnitude than that contained in our entire universe, or the cause is something that is not energy as we know it but which can create (God-like) the energy we observe. (See Element 1)

Element 5: *What we perceive as solid matter is almost entirely empty (of matter) space.*

The structure of matter at the subatomic level (the so called quantum level) is so complex that science is still trying to decipher it. However, at the atomic level the structure of matter appears to be well defined. An atom of any element contains entities called electrons, protons and neutrons. That protons and neutrons (electrons?) are themselves made up of other entities does not alter the following facts.

Many calculations and measurements have confirmed that protons and neutrons behave much like particles that clump together in what is referred to as the nucleus of the atom. The electrons, however, behave more like an energy field, surrounding the protons and neutrons like a large (relative to the nucleus) cloud. But this cloud contains virtually no matter. In round numbers it is found that the space occupied by the electron cloud and the nucleus encompass a volume that's 10,000 times greater than the combined volume occupied by the nucleus and electron as particles. Stated another way, the volume of matter in the atom is only 0.01% of the total volume occupied by that atom, which means that the atom is 99.99% empty of matter. You, the chair, the earth, everything that is perceived as matter – solid, liquid or gas – are at least 99.99% empty (of matter). It is the interlocking energy fields of atoms that impart structure to matter.

Element 5 is an uncontested scientific fact verified by many physical measurements.

Element 6: *Empty (of matter) space is filled with energy.*

Empty (of matter) space, whether internal to individual atoms, in the space between atoms or everywhere external to atoms is filled with various forms of energy. Some of this energy is known and can be measured. A partial list of such energy includes electrostatic fields, magnetic fields, the electromagnetic spectrum (visible light, infrared, ultraviolet, x-rays, gamma rays, etc.) and a wide array of subatomic entities that pop into and out of measurable existence and behave either like particles or energy waves, depending on how they are measured.

In addition to these known energy forms there are theoretical and experimental results that suggest that the entire universe is immersed in a uniform energy "field" of currently unknown nature. The problem with a uniform field that exists everywhere is that there is no point of reference for measuring its existence; it must be inferred in some manner. This is not an easy problem to solve, if solvable at all.

In any event, Element 6 is an uncontested scientific truth, with or without the addition of a uniform energy field.

Element 7: *The passage of time is not the same everywhere.*

Prior to Einstein, the passage of time was thought to be the same for everyone. But Einstein derived equations that predicted the passage of time to be relative to the speed of travel through space. The concept of "space-time" was born and our perception of how the physical universe operates was forever changed. The space-time concept, as structured from Einstein's Special and General Relativity Equations, has been proven in all tests to date. In fact, it is necessary to use the relativity equations to make our global positioning system (GPS) work.

It is now known that the passage of time slows (relative to someone who is stationary) the faster one is traveling through space. As an example, if you could leave earth for a one year (your time) round trip into outer space, traveling at 99.99995 percent of light speed, 3000 thousand years will have passed for those on earth. (Not my calculation; drawn from an example in "The theory of Almost Everything" by Robert Oerter.)

It would be interesting to make such calculations for an observer traveling with the "big bang" expansion of the universe to see what time would have elapsed for this observer — relative to our time — from start to present.

In any event, Element 7 is an established fact reduced to practice in many technical calculations.

Element 8: *All life is initiated by, and dependent upon, the consumption of energy.*

However life was initiated, in whatever form, energy input was required to organize inanimate matter (itself a form of energy) to the higher energy state required for life. This life is then dependent upon obtaining sufficient energy to maintain life's processes. If sufficient energy is not found when needed, or the organism is otherwise sufficiently damaged, life processes cease and the organism dies.

In whatever manner an organism dies, the energy associated with its life is released back into the surroundings (energy is neither created nor destroyed). Other organisms compete to use some of this energy by consuming (in one form or another) the dead organism in order to maintain their own life processes. These other organisms also eventually die and release whatever life energy they have accumulated back into the surroundings (including to other organisms) and so on. In this manner the so-called "circle of life" is established, fueled (primarily) by the radiation energy received from the sun while functioning (primarily) through organisms consuming the energy other organisms.

The above description of the central role of energy in life is so well documented that no further proof will be offered here. The source of the energy that initially created life may be logically unknown but the necessity of consuming energy to create and sustain life is unquestionable.

Element 9: *The evolution of all life forms is an established fact but the force(s) driving evolution remain unresolved.*

That all life forms evolve is no longer scientifically or logically uncertain. The ability of rapidly replicating viruses and germs to develop resistance to previously deadly chemicals is one directly observable proof of this. Selective breeding of plants and animals to produce desired traits is nothing more than manipulating genetic codes, which is the same mechanism driving evolution. The discovery and mapping of genetic codes graphically shows the progression of evolution over long periods of time. In short, all life continually evolves, unless it fails to adapt to its environment and dies.

However, proof that life forms evolve does not prove that "evolution", as popularly advanced, is true. The popular conception of evolution is that it is the result of randomly caused genetic mutations that enable an organism to better adapt to its environment. The causative agent is assumed to be one or more of many possibilities from spontaneous genetic rearrangements through chemical interventions to interactions with high-energy particles. Advantageous mutations are those that permit an organism to better survive in its environment, thereby creating more offspring.

In fact, all of the causative agents described above, and more, can be demonstrated to cause genetic mutations. These causative agents have always existed and are still at work in the world today. It can also be unequivocally observed that organisms do successfully adapt to their environments in order to survive while others fail to adapt and go extinct. What remains in contention is the assumption that these causative agents, <u>randomly encountered</u>, explain both the creation of life from inanimate matter and the awe-inspiring variety of life forms, past and present.

The role that randomness plays in evolution is far from logically established. If something is truly random then it obeys statistical probabilities. If some mutation, or string of mutations, is not statistically probable then they are not random. The calculations to establish the statistical probabilities for mutations require knowledge of the stepwise sequence of molecular changes occurring. Until recently this knowledge was not available. As a result of this ignorance the proponents of strict random mutation as the driving force behind evolution were given a free ride for their claims. That free ride is ending.

With the advances occurring in molecular biology, tracking the actual molecular changes occurring in a particular mutation is becoming feasible. Statistical probability calculations of mutations involving simple life forms (the only studies now feasible) do not support randomness when the complexity of required molecular changes increases beyond some threshold. These calculations consider the number of possible outcomes resulting from a random molecular event (bond breaking, etc.), the probability of such an event occurring, the number of possible results from this event and the time available for this event to produce an evolutionally helpful mutation. There is no other way to logically resolve the role of randomness in producing complex life forms.

The proponents of randomness have yet to accept such statistical conclusions but offer no calculations of their own to support their position. Instead, they simply assert that evolutionary change is observable and that nothing more than random chance acting over time is needed to explain such evolution. This is not logic but an unsubstantiated belief.

As the tools of molecular biology improve, the role of randomness in evolution will eventually be open to rigorous scientific analysis. At present, evolution is a fact but what fundamentally drives it, and in what manner, remains unresolved.

Derivative Elements of Physical Reality

Basically, all of the discoveries of science and engineering may be considered Derivative Elements of the Fundamental Elements of Physical Reality (as we know it today). I will make no attempt to reference such discoveries beyond what I have done in the preceding discussions. There are simply too many of them, although the details can be fascinating. There is only one Physical Derivative Element I feel warrants definition because of the perspective it provides in regards to Man's scientific understanding of the universe.

Physical Derivative Element 1: *All of Man's understanding of the universe is limited to the "how" not the "why" of its operation.*

Why the universe operates as it does is forever closed to scientific understanding because the creation of the universe is closed to scientific understanding (See Fundamental Physical Element 1.) The fundamental laws and properties of the universe are part of Creation and are thus inseparable from the God Power for their existence. Man, at best, is but an observer of how the universe functions, hoping to utilize what is learned for the betterment of Mankind and the world in total.

Unfortunately, what is learned can an also be utilized for less noble objectives.

I submit the preceding nine Fundamental Elements of Physical Reality and one Derivative Element as the core of physical reality. These ten elements provide a firm foundation for basing one's perception of physical reality. They also provide the first criteria to which the following Elements of Psychological Reality must conform to if they are to be considered true elements.

Elements of Psychological Reality

(In what follows the term 'behavior" encompasses all of a person's mental/emotional/physical responses to a stimulus/environment. The term "action" is reserved for the subset of those responses that can be directly observed.)

Practical Definition of Psychology

A search for the meaning of the terms "psychology" and "psychological" produces a range of possible definitions. The dictionary definition that best fits our use herein is: "the science of mind and behavior as applied to humans". Unfortunately, this definition has little practical use other than to suggest that mind and behavior are associated.

A more useful (albeit longer) definition of psychology can be stated as: "Psychology is the mental state created by the interaction of a persons <u>modified</u> nature with his/her <u>perceived</u> reality. A modified nature (assuming there are no physical impairments involved) is the inherited nature modified by inputs from acquired beliefs, life experiences, knowledge and reasoning. The perceived reality is constructed by these same inputs in association with the modified nature. This mental state (psychology) dictates behaviors/actions consistent with the perceived reality."

This definition of psychology identifies the five factors - inherited nature, beliefs, experience, knowledge and reasoning - which shape our psychology and ultimately dictates all our behaviors/actions.

The concept of psychology is only applicable to the individual as the individual is the source of all human actions. When joined with others to form a group the "actions" of this group are simply the accumulated actions of the associated individuals. There is no group psychology separate from the individual psychologies involved. Therefore, to understand human behavior in general one must understand the behavior of the individual.

Overview of Psychological Reality

Before dealing with the elements of psychological reality an overview of human behavior is useful for setting the scope for what is involved. When examined in detail human behavior is quite complex but one need not get lost in this complexity to gain a practical understanding of its core. To describe this core, I offer the following:

The individual acts on the world around him/her in keeping with that individual's psychology *(modified nature interacting with his/her perceived reality)* responding to a stimulus. This function of psychology makes it a critical factor in all human interactions. For any stimulus the logic is as follows:

Claim 1: The prevailing psychology of the individual creates the impetus for action.

Claim 2: It is actions, not thoughts nor intentions, which directly produce consequences.

Conclusion: To produce desired actions, and thus desired consequences, the appropriate psychology must be active.

Overly simplistic as it may seem, this logic sequence captures the essence of all human behavior. The validity of the logic depends only on the validity of Claim 1. Claim 2 and the conclusion are logically unassailable if Claim 1 is true. Therefore, to determine the validity of the overall logic sequence one needs only to verify the role that psychology plays in dictating actions. To do this – as psychology is a nebulous state of mind – there is no practical choice except to work with the factors that create this state – one's modified nature interacting with one's perceived reality.

The five factors that create our modified nature and perceived reality, thereby dictating our actions, are inherited nature, beliefs, experience, knowledge and reasoning. The role each factor plays, and the susceptibility of each to manipulation, can be summarized as follows:

Inherited nature is the most complex factor in that it encompasses all the intellectual, instinctual and emotional characteristics that humans possess at birth. Human nature is the raw material the expression of which the other four factors modify to varying degrees. At birth no single characteristic is expressed equally in all people but sufficient similarity exists that human nature is frequently assumed to be universally shared. However, this is an assumption with many exceptions, some apparently random in nature, others the result of genetic inheritance.

The initial plan for one's nature is set at conception via one's genes. The gestation period following conception functions, with varying success, to carry out this plan. The characteristics of human nature resulting from conception and gestation are subjected to modification immediately following birth (perhaps before birth) through inputs from the other four psychological factors. This modification acts to suppress or enhance the expression of one's inherited characteristics; it does not remove or instill new ones (except, perhaps, in evolutionary time frames).

Some characteristics of human nature – such as pain, fear, hunger and sex – serve primarily to promote personal and/or species survival. As such these so-called "primal instincts" are strong and not easily resisted. Exploiting the primal instincts provides the most direct and frequently employed route for influencing the actions of others (think "do or die" at the extreme).

Beliefs range from simple concepts, through logical convictions, to complex philosophical constructions ultimately requiring faith for acceptance. Beliefs are not with us at birth but must be learned and accepted. One's beliefs are constructed with input from all the other factors at work in human psychology but often in ways more subconsciously than consciously acquired. Beliefs, once acquired and embraced, can become the most influential of the psychological factors in that they can justify any action and need not be true nor morally/ethically "good". Creating, instilling and manipulating beliefs have historically proven to be a frequent and very effective means for influencing the actions of others.

Experience is always a factor in one's psychology but how powerful it is in relation to the other factors depends upon the intensity and immediacy of the experience. Particularly intense experiences, good or bad, can become an intellectual and/or emotional benchmark against which all subsequent psychological inputs are compared for relevance. However, the problem with many experiences is that while the outcome appears clear (you experience it) the cause may have multiple, unrecognized sources. As a result, wrong conclusions are common as to what an experience teaches, but these conclusions are strongly held. Advancing false causes for experiences is a proven way to influence the actions of others.

Knowledge functions as raw material for the reasoning factor. Limit the supply of knowledge and what can be deduced through reasoning is likewise limited. Feed in false information masquerading as truth and the reasoning process results in false conclusions, which conclusions cannot be recognized as false until true information replaces the false. Limiting the dissemination of true information and spreading false information is very effective in influencing the actions of others. Combining the dissemination of false information with controlling the education process is today's preferred "enlightened" method of influencing the actions of others.

Reasoning acts to balance the inputs from the other four factors so as to rationalize a reality that, in turn, dictates actions consistent with that reality. The balancing can be as straightforward as the afore mentioned "do it or die" to as complex as assessing the meaning of life. The reasoning process is vulnerable to all kinds of defects that can lead to wrong conclusions. The inability to separate emotions from logic, focusing on the immediate at the expense of the longer term, or simply not making the effort to apply logic to a situation are three very common defects. But the most damaging defect is simply not working with the truth. Reasoning with wrong information always, barring compensating errors, leads to wrong conclusions, ultimately resulting in wrong actions.

The conclusions, true or false, reached through reasoning not only guide subsequent actions but also feed back into the other four psychological factors to further modify one's inherited nature and perceived reality, thereby dictating new actions consistent with the changes made.

Thus, the reasoning factor in human psychology is slave, moderator and dictator all in one. It is a slave to the input of the other factors but can feedback to modify those same inputs, and it is a dictator of our actions. Reasoning does all this through the modified nature and perceived reality it constructs. **Thus, the key to controlling the actions of others lies in influencing their modified nature and perceived reality.**

Of course, this is also the key for controlling our own actions.

Psychological Element Constraints

Efforts to understand the mind/behavior relationship (psychology) are as old as mankind itself. Throughout history intellects great and small have observed, recorded and pondered man's behavior in various situations in order to understand this relationship. Countless such studies and observations over time have produced a body of knowledge relating to man's psychology that is extensive and well documented. Also well documented – man's nature being what it is – are the methods that have proven effective at manipulating psychology.

This recorded psychological knowledge is available for study by anyone in a free society. But, as you might suspect, it is not easy to sort through this record to identify useful information. The definition of "useful" in this context means useful for understanding, influencing and predicting human actions. The Elements of Psychological Reality provide the structure on which useful information rests.

The context from which true elements of psychological elements can be drawn is bounded by two constraints:

Constraint 1): Only long-term, historically consistent observations/interpretations of psychological truths can be trusted. Psychology – "the science of mind and behavior" (an alternate Webster dictionary definition) – is not a precise science. The "mind" is not the physical brain but the thoughts that the brain produces. As such, the mind has no properties that are directly measurable. Contrary to what some propose, physical mapping of the brain to locate areas where memories are stored or various functions are performed is not measuring the mind.

The science part of psychology consists solely of observing and interpreting associations between a stimulus and the resulting behavior that the stimulus produces. However, total objectivity in associating stimulus with behavior has proven difficult to attain, resulting in conflicting explanations of what constitutes psychological "truth". Hence, there is need for many such stimulus/observations/interpretations over time to establish confidence in the results.

Constraint 2): All Fundamental Psychological Elements of Reality are derived from the psychology of the individual. There is no "group psychology" independent of the individual that produces actions; there are only the accumulated actions of the individuals involved in the group. Individuals in a group can be influenced by one another but only when the psychology of each individual permits it. Therefore, all psychological elements necessarily pertain to the individual.

As was previously done with the elements of physical reality, key fundamental elements of psychological reality will first be listed without justification. This approach provides a relatively uncluttered view of psychological reality as a whole (assuming the elements listed are accepted as true). The reasoning for including each element as true will then follow, element by element. As the truth of each element is established the structure of psychological reality thus revealed must fit the world as it is experienced. If it does not fit the world as experienced then at least some elements are in error and the reality described is false.

Psychological Fundamental Elements

It is the fundamental elements of psychological reality that set the stage for all the actions of mankind. There are just five such elements:

1. All individuals are born with a common set of characteristics/instincts.

2. At birth no inherited characteristic/instinct is expressed equally in all people. The intensity of expression for any individual characteristic or instinct can range from extremely weak to obsessively strong.

3. The expression of an individual's inherited characteristics/instincts (nature) is subjected to modification immediately following birth (perhaps before birth) through the interaction of four inputs: acquired beliefs, experience, knowledge and reasoning.

4. An individual's perceived reality is constructed from the interaction of five inputs: modified nature, beliefs, experience, knowledge and reasoning.

5. Except in cases of physical brain defects the individual always behaves/acts in accordance with the dictates of his/her nature and perceived reality.

To establish the validity of the individual elements so as to justify their status as True, I offer the following rationale:

Element 1: *All of humanity is born with a common set of characteristics/instincts.*

Consider a typical (albeit partial) listing of man's characteristics/instincts as identified over the course of man's existence:

- **Hate** (consuming, passionate dislike of someone or something)
- **Self-preservation** (desire to protect one's life)
- **Love** (includes a wide range of physical, emotional and spiritual connections to self and others)
- **Compassion** (sympathetic consciousness of others' distress)
- **Greed** (desire to accumulate something to excess)
- **Pride** (self-esteem)
- **Arrogance** (an attitude of superiority)
- **Envy** (resentment of what others possess)
- **Dominance** (desire for power over others)
- **Aggression** (the use of force, actual or implied, to obtain one's desires)
- **Sex** (the drive to propagate one's species)
- **Lust** (destructive sexual desires/actions)
- **Intellect** (learning/reasoning ability)
- **Curiosity** (desire to understand/experience)

Each bold script term in the list above is intended to represent a characteristic/instinct influencing man's behavior. By themselves such terms encompass so many shades of meanings that they have limited value in communications. Hence, to improve meaning definitions must be attached to each term, such as those attached in parenthesis above. As listed, these terms and definitions are widely accepted (in English) as representing key aspects of man's inherited characteristics/instincts. However, different cultures frequently define and interpret them somewhat differently, sometimes greatly different. Also, while the terms as defined have proven useful for conceptualizing the forces driving man's behavior one must always bear in mind that no single characteristic/instinct acts alone. The interactions among the various characteristics/instincts are many and varied.

All of recorded history confirms that mankind shares these characteristics/instincts. From the earliest "survival of the fittest" clans to what we today call "enlightened" modern societies — including all races and ethnic groups regardless of the societal environment or level of knowledge in which they exist — all demonstrate the consequences of this basic nature. Such universality of human nature emanating from such diverse environments precludes it from being a "learned" condition; only an inherited condition encompasses such diversity.

Additional, more specific evidence supporting inheritability comes from studies involving identical twins that are separated at birth, raised in different environments and having no contact with each other while growing up. In such cases studies invariably find that the twins develop remarkably similar behaviors and capabilities, even to the extent of developing identical subconscious mannerisms.

The foregoing does not mean that the <u>expression</u> of man's inherited nature is the same for all mankind. The evidence only establishes that man's <u>basic</u> nature is universally inherited.

Element 2: *At birth no inherited characteristic or instinct is expressed equally in all people.*

The evidence supporting this statement is compiled by observing the behavior of people as they grow. Even between newborns there can be noticeable

45

behavioral differences. Some newborns are timid, some bold, some outgoing, some reserved, etc. As children grow various other behavior differences surface in temperament, aptitude, emotional behavior, intellect, etc. – as any mother of multiple children can attest. To be sure, one's environment when growing up is a significant factor in how an inherited nature is expressed but the environment is never completely dominant. History abounds with examples of people overcoming terrible environments or, conversely, going "bad" even when they enjoy the best of environments.

Element 3: *The expression of an individual's inherited characteristics/instincts is subjected to modification immediately following birth (perhaps before birth) through the interaction of four inputs: acquired beliefs, experience, knowledge and reasoning.*

It is self-evident that a major part of growing up includes learning to adapt to one's perceived reality. This adaptation includes suppressing or enhancing the expression of the various inherited characteristics/instincts to better fit the perceived reality. The process of adapting continues throughout life in either a planned proactive manner or through unplanned reactions to the consequences of one's environment or behavior. In all cases the modification is driven by the perceived reality.

Shaping this perceived reality to enhance or suppress inherited characteristics/instincts is the role that education (beliefs, knowledge, reasoning and experience) play in an individual's life. While the individual retains ultimate control of whether an inherited characteristic/instinct is enhanced or repressed, everyone is susceptible to such modification.

Unsurprisingly, the stronger the inherited characteristic or instinct the more difficult it is to modify. Both history and current everyday experience provide many examples of individuals expressing extreme inheritable characteristics or instincts through their actions. Such actions, good or bad, can be judged extreme if they are clearly recognizable as outside the expected norm. News reporting depends upon such extreme examples of actions for much of their appeal. There is no shortage of examples on a worldwide basis.

(The author recognizes that the term "extreme" is relative to what one considers "normal" or, perhaps more accurately, what one considers "desirable". However, this complication does not lessen the validity of Element 3.)

Element 4: *An individual's perceived reality is constructed from five inputs: nature, beliefs, experience, knowledge and reasoning ability.*

Element 4 is true simply because the five inputs identified encompass all possible inputs -- rational or irrational -- involved in forming a view of reality. How these inputs act and interact in constructing a perceived reality was outlined earlier and need not be repeated here. Corrupt any of these inputs with untruths and the resulting view of reality will be false.

Element 5: *Except in cases of physical brain defects the behavior/actions of an individual are always in accordance with the dictates of his/her* *expressed (modified) nature interacting with his/her perceived reality.*

Like Element 4, Element 5 is true because all the inputs that can guide or trigger actions are encompassed by the interaction of one's expressed nature with one's perceived reality. This interaction constitutes one's psychology. Recognize this truth and all the actions of mankind become logical consequences of such interaction.

The expressed nature may not be logical and/or "good", and the perceived reality may be false, but this is all the information the individual has on which to base an action. Actions may be the result of careful planning, a spontaneous response or conditioned reflex to some stimulus. It makes no difference how the action came to be. One's expressed nature and perceived reality dictated the action.

The preceding five Fundamental Elements of Psychological Reality form the basis for all human actions and serve to explain such actions (largely in hindsight). But, they convey no information as to the probability of a specific future action occurring from among all those possible. This is where the long-term observation of human behavior comes into play. Such experience allows one to estimate the probability of a particular action occurring in specific situations.

Unfortunately, even with experience estimating the probability of action with respect to a randomly chosen individual remains a difficult proposition. However, with respect to a sufficiently large group of individuals – all such individuals randomly chosen – experience frequently does allow for quite reliable predictions. These predictions take the form of Derivative Elements of Psychological Reality, which we will now address

Psychological Derivative Elements

The consequences of the Fundamental Psychological Elements dictate that any group of individuals will contain a mix of expressed psychological traits. In any such mix there is a probability that a particular expressed trait – from among all possible expressed traits – will be present in one or more individuals. Should these probabilities be known then one could make statistical predictions as to how the group, or someone(s) in the group, will act in various situations. The problem is that the probabilities needed for random groups are seldom reliably known, even when education, propaganda, lies etc. have influenced the outcome.

There is one exception to this problem of reliability and that is where experience shows the probability to be 100%. There is no estimating involved in determining whether a particular action occurs with 100% certainty – it either does or does not always occur (as verified by long-term experience).

Because true elements must remain true in all instances within their scope, Derivative Elements only exist for situations with 100% certain outcomes. This does not mean that estimated probabilities of less than 100% are without value, just that there is always a risk of being wrong with such estimates. Such uncertainty would preclude making clear true/false assessments of Derivative Psychological Elements.

The six Derivative Elements in the following list were selected as being particularly pertinent in today's world. History has proven each to be 100% certain and each has significant consequences in both personal and world affairs.

1. In any sufficiently large population of people an extreme expression of all inheritable characteristics/instincts will be found to exist.

2. If a situation exists that can be exploited to someone's advantage, that situation will ultimately be so exploited.

3. All organizations/associations — organized religions included — are in operation social constructs subject to all the strengths and weaknesses of the individuals involved.

4. Shaping the perceived reality is the central battle between competing interests.

5. Some competing perceived realities <u>cannot</u> forever peacefully coexist.

6. Violence always has and always will play a role in man's interactions, employed by some for advantage, required by others for defense.

The truth of these Derivative Elements is based upon the following evidence, verified by long-term experience:

Element 1: *In any sufficiently large random population of people an extreme expression of all inheritable characteristics/instincts will be found to exist.*

Fundament Psychological Element 3, as previously reviewed, has established that extreme expressions of all inherited instincts/characteristics) exist in the world. For establishing the truth of Derivative Element 1 the only remaining issue is what constitutes a "sufficiently large" population.

To resolve this issue one turns to the record of human experience.

As with all statistical estimates the sample size (population) needed for certainty depends upon the nature of the population involved, the particular attribute considered, and, sometimes, the time span allowed for observation. As history documents – and current news is eager to report – an extreme expression of all inherited instincts/characteristics manifests itself somewhere in the world virtually every day. The world population is, of course, the ultimate in a "sufficiently sized" random population.

Closer to home, the extremes of most instincts/characteristics are also evident in any large city. While the dynamics of large cities do introduce biases that favor some characteristics or instincts over others, apparently there exists sufficient sample size and randomness to ensure that all extremes, over relatively short time frames, are demonstrated. For proof one needs only to review the news of record for any large city such as Chicago, New York, Los Angeles, etc. A review covering a time span of only six to nine months is probably sufficient.

For smaller cities the reduced sample size increases the chance for biases to significantly impact randomness but it appears that smaller cities also experience extremes in instincts /characteristics, it just requires longer time frames for all extremes to occur.

Actually, in today's world the mobility of people is such that it is best to initially assume that any group of people may include, at some point in time, one or more persons harboring extreme instincts/ characteristics (good or bad). Of course, the probability of such extremes existing or not existing in a small group, at a specific time, depends heavily on why the group was formed.

Element 2: *If a situation exists that can be exploited to someone's advantage, that situation will ultimately be so exploited.*

Results of this element at work in the world are everywhere evident. Exploitation ranges from the benevolent to the evil with all gradations in between. Benevolent exploitation is usually encouraged, such as an entrepreneur seeking personal gain by filling a patent for a new product. The obviously evil or malicious exploitations are usually discouraged by penalties established through laws and regulations. In between these extremes are many unethical or immoral exploitations that do not quite meet the evil or malicious threshold but which nevertheless degrade human interactions.

While virtually everyone recognizes that unwanted exploitations are all around us, not everyone explicitly accepts the truth that such exploitation is an eventual certainty wherever the opportunity exists. But the history of man's affairs demonstrates time and again that where an opportunity for exploitation exists they eventually will be exploited. By themselves the multitude of laws and regulations that man has enacted attest to the pervasiveness of exploitations. (Unfortunately, history also shows that governments frequently use laws and regulations for their own exploitive purposes.)

The key to minimizing unwanted exploitations is to embrace the truth of Derivative Element 2 and take objective, proactive steps to reduce the <u>opportunities</u> for unwanted exploitations. Penalties after the fact have their place but reducing the opportunity for exploitation is by far the least painful and more effective deterrent. How this works in practice will be explored later via examples in everyday life.

Element 3: *All organizations/associations – organized religions included – are in operation social constructs subject to all the strengths and weaknesses of the individuals involved.*

This derivative is a consequence of the Constraint #2 of psychological reality, which recognizes that all actions of mankind are the result of individual actions, whether carried out by one individual from beginning to end or by many individuals, each having a part in producing the action.

All organizations or associations have a purpose for existing whether that purpose is explicitly stated and of long duration (i.e. governments, businesses, religions, etc.), or simply an instantaneous or short term gathering of liked minded people. Whether an organization or association is good or bad is determined by the actions of the individuals involved, regardless of the stated purpose or intentions of the organization.

The stated purpose (if there is one) of an organization or association is often of little relevance to how the individuals involved actually perform. History clearly documents how easily even the loftiest of purposes are corrupted or negated to benefit those in control.

Element 4: *Shaping the perceived reality is the central battle between competing interests.*

Recall that in the absence of physical brain defects the interaction of an individual's perceived reality with his/her modified nature provides the impetus for all human action. The nature an individual inherits at birth cannot, in general, be influenced before birth. However, following birth this nature undergoes more or less continuous modification. Such modification is accomplished through the perceived reality the individual embraces.

One's perceived reality is constructed with the five inputs identified earlier: nature, beliefs, experience, knowledge and reasoning. The last four of these inputs can be directly influenced by others and by so doing indirectly modify an individuals inherited nature. The result is a perceived reality and a modified nature that can, if correctly manipulated, produce actions beneficial to whomever is doing the manipulation.

For the relatively short term (which history shows can be hundreds of years long) those in power have shaped the perceived reality of others through a combination of brute force – to remove non-believers and to convince any survivors that resistance is futile – and by establishing a system of beliefs that justifies their use of force and right to rule. This approach is still used in the world today but is becoming increasingly costly and difficult to sustain on a large scale, in part because the communication revolution is exposing more individuals to alternate realities (hopefully, true realities).

In the "enlightened" societies of today the primary methods used to shape the individual's perceived reality are through control of the education process and the flow of information available to the individual. These methods are the same whether trying to establish false or true perceived realities: false information for false realities, true information for true realities. Successfully done this approach is less costly and less risky for those seeking to shape the perceived reality. This approach also yields longer-term success. Direct force is still judicially used where deemed necessary but such force is usually filtered through laws, rules and regulations.

Of course, in the less "enlightened" societies around the world blatant brutality remains the tool of choice.

Element 5: *Some competing perceived realities cannot forever peacefully coexist.*

The truth of this derivative is obvious in situations where people holding one perceived reality are actively trying to eliminate people holding a different perceived reality. However, this truth is not always so obvious when active hostilities are not underway. The world is full of perceived realities that fundamentally conflict on issues vital to the individual, issues such as personal freedom, right to life, rule of law, the power of government, etc. Most of the long and bloody conflicts in mankind's history revolve around these vital issues.

It is rational, not paranoid, to carefully analyze the elements, true or false, on which various proposed realities are based. Only in this way can one fully understand the true source of present and probable future conflicts and effectively initiate solutions (or defenses).

Element 6: *Violence always has and always will play a role in man's interactions, employed by some for advantage, required by others for defense.*

This truth is obvious to anyone who knows history or observes events in the world today. Man's nature and the existence of opposing perceived realities make Element 6 inevitable. The best that can be done to minimize violence is to integrate the elements of true reality into everyone's perceived realities. With perceived realties based on True Reality the majority of people will find common interest with one another. Unfortunately, some individuals will always see violence as an effective tool for personal gain and/or power over others.

So it has been throughout history.

Elements of Transcending Reality

Background Summary

Fundamental and Derivative elements explain much of how individuals and societies operate, particularly when man's "survival of the fittest" instincts are unrestrained. Unfortunately, when such instincts are unrestrained man's record of behavior toward other men, indeed toward the world as a whole, has not been good. Men and woman of good will have long sought solutions to this problem. Such effort has resulted in an assortment of truths that I group under the heading of "Transcending Elements".

Transcending Elements were initially articulated and advanced through religious and philosophical arguments that were not widely available to mankind as a whole. Nor was mankind — conditioned through millennia by the "survival of the fittest" psychology, servitude to the strongest and general ignorance — eagerly receptive of such propositions.

Further, Transcending Elements all deal with the betterment of mankind by essentially improving and empowering the individual. Thus, these elements were, and still are, resisted by prevailing power structures that by their nature do not like their power usurped by the individual. The result of these various factors is that Transcending Elements are always under attack — either through force or through the promotion of false realities — by those seeking power over others.

Today the situation with regard to attacks on Transcending Elements is only marginally improved over that in the past. Man's knowledge of physical reality is vastly greater today and the systematic study of man's psychology has resulted in a better understanding of why we behave as we do. These advances create the opportunity for the individual to form a more complete vision of reality. On the other hand, man's inherited nature has changed not at all, with all of man's inherent strengths and weaknesses still active and vulnerable to false realities. For those seeking power this creates the opportunity to apply modern technology and knowledge to more effectively promote false realities. Transcending Elements are the hard earned truths that false reality proponents most fear.

Listed below are truths that form the core of Transcending Elements. Because Transcending Elements are such critical components in an individual's perceived reality, great care must be exercised in both validating and protecting them from corruption.

Transcending Elements

Element 1: *The God Power created "Everything" for a purpose; the task for Man is to identify and fulfill this purpose.*

We will never know — through our own efforts -- the ultimate purpose behind the creation of "Everything", for to do so would require knowledge on a par with that of the God Power. Science provides us with glimpses as to how the universe has operated after its creation but can provide no insight as to what overall purpose it serves. Religions propose various scenarios for the purpose behind creation but such proposals sometimes deviate from the dictates of true reality. The best Man can do is to build our knowledge on the foundation of True Reality and see where that leads us.

"Evolution" appears to be the God Power's method for creating and developing life forms. Evolution is a hard taskmaster in that if a life form does not serve a purpose in the changing web of life it does not long exist (at least in geologic time frames). Man is the first life form with the intellectual ability to contemplate a purpose for its existence and to recognize that there must be such a purpose. Consequently, Man has struggled with this question of purpose probably from the beginning of cognitive thought.

Perhaps this question of purpose is not so difficult to answer when approached within the context of True Reality. We will never know for certain we have identified the God Power's intended purpose but an identified purpose that meets all the dictates of True Reality is unlikely to be far off the mark.

Element 2: *Any God Power relationships with humanity are with the individual.*

This element may seem so logically obvious as to not warrant listing as a Transcending Element. After all, only physical sentient beings can enter into such a relationship. Groups or organizations comprised of individuals create no physical sentient identity separate from the members themselves. They do create leaders that (hopefully) represent the consensus views of the individuals they lead. However, no individual – leader or otherwise – can know the mind of another individual so completely as to represent him/her before the God Power. Thus, all God Power relationships are necessarily with the individual.

The reason this truth is included as a Transcending Element is because it identifies the individual as the key link in the God Power's plan for humanity. It is through the individual that the plan for all of mankind is being implemented. It is up to the individual to decide -- through the exercise of their "free will" that his/her intelligence permits - what this God Power plan is and how best to follow it.

Element 3: *The individual is the template that mankind follows.*

Transcending Element 3 is a consequence of the fact that the individual is the source of all human actions. There are no group actions separate from the cumulative actions of the individuals involved. Each individual's actions are determined by the interaction of his/her modified nature and perceived reality. Thus, the prevailing modified nature and perceived reality of individual members -- taken in aggregate -- becomes the template that the group reflects. This relationship holds true for any group with three or more members, up to the whole of humanity.

Differing modified natures and perceived realities gives rise to groups with differing objectives, which objectives sometimes conflict. *There is no long-term way to avoid these conflicts except through harmonizing the modified nature and perceived reality of all involved with True Reality.*

Element 4: *Only perceived realities founded on True Reality can create the conditions for a better world.*

In essence, True Reality is a glimpse into how the God Power is implementing the evolution of "Everything". Man cannot thwart the God Power's dictates. The best we can do, using all information at our disposal, is to seek harmony with what we understand as True Reality and its dictates. The maxim "Man proposes, God disposes" is an accurate summary of Man's efforts in the world. Only by following the God Power's plan, the best estimate of which is based upon True Reality, can Man hope to make progress in bettering the world.

There is only one True Reality, the components of which are not all known. However, we do know many such components and any proposed reality that does not include them is a false reality. It is the individual's responsibility to seek out and embrace true reality. False realities never lead to beneficial outcomes in the long term, and rarely so in the short term.

Element 5: *Only the individual can create a better world and he/she can do so only by adhering to the dictates of True Reality.*

This Element is the logical summation of the preceding Elements 3 and 4 but more clearly states the role of the individual in world affairs. Of course, the more individuals with allegiance to True Reality the bigger their impact on world affairs will be. There is no path to improving the world except through the dictates of True Reality and there is no agent to traverse this path except the individual. It is particularly important that those individuals in power adhere to the dictates of True Reality.

Those promoting false realities well know the truth of Element 5, which is why they will go to any lengths to confuse the individual as to what constitutes True Reality. For their purposes confusion works almost as well as convincing an individual to accept their false version of reality.

Uncertainty as to what is true makes the individual more willing to try something that sounds desirable even if offered without proof of validity. However, whatever the promises or realities that men (or women) make or promote, however desirable they may seem, all are certain to be detrimental to mankind if they are not in compliance with True Reality.

Element 6: *Just four attributes – Love, Honesty, Humility and Responsibility – of an individual's modified nature are needed to achieve maximum harmony in the world. (Assuming True Reality is the perceived reality).*

The basis for maximum harmony in the world is encapsulated in the words "*We hold these truths to be self-evident, that all men are created equal, that they are endowed by their Creator with certain inalienable rights, that among these are Life,*

Liberty and the pursuit of happiness". For conciseness and substance these words cannot be improved upon. If we are to progress toward the goal of maximum harmony the individual's inherited nature must be modified to support these inalienable rights.

As was discussed previously a modified nature, assuming no physical impairments, is the inherited nature modified by inputs from acquired beliefs, life experiences, knowledge and reasoning processes. The interactions between the various inputs and one's inherited nature are many, complex and, in total, unique for each individual. However, for practical purposes the attributes of a modified nature that support maximum harmony in the world are not that many, albeit each is complex and challenging both to acquire and to maintain. There are four such attributes. The terms (in English) used to represent them are:
Love,
Honesty,
Humility
and Responsibility.

Anyone researching the meaning of these terms – via dictionary, philosophical or theological sources – will find many interpretations to consider. One obvious finding is that there are significant differences among the various interpretations. Thus, in order to avoid misunderstandings when dealing with these terms one must know which interpretation is in play. Such knowledge is not always obvious or easily obtained, particularly when differing religions, ethnic groups or deceit is involved. When encountering these terms it is best to never assume that your interpretation is the one that everyone involved endorses. If, in a specific situation, an unambiguous understanding of what the term encompasses is important then a written statement of what the term encompasses is required for all involved to review. It is impressive what disagreements such a written statement can expose when all involved thought the meaning of the term was clear.

While the above terms are all attributes of the human mind they ultimately manifest themselves as actions through interaction with one's perceived reality. Assuming that the perceived reality closely approximates True Reality then these four attributes largely determine how beneficial such actions will be. It is crucial to understand what these four attributes encompass in the context of world harmony. I offer the following abbreviated interpretations, focusing only on those actions that can be clearly linked to <u>universally</u> beneficial consequences in Man's relationships.

Love is the most powerful of all attributes. It can be convincingly argued that love, in one form or another, is an integral part of all beneficial attributes of human nature. As applied to humanity as a whole the essence of love is summed up in the Christian directive to "Love your neighbor as yourself". All existing great religions have similar directives but not all apply them universally. True universal love requires that you accord to everyone, at their creation, the "inalienable rights of life, liberty and the pursuit of happiness". It also impels one to assist others to achieve these rights with the goal of propagating universal love to the maximum extent possible.

However, universal love is not blind. It recognizes that, while such love may be offered to everyone, not everyone will immediately (or ever) accept this concept, let alone reciprocate or offer it to others.

Thus, measures must be taken to protect one's self (or others) while maintaining the offer of universal love, even for to those who oppose it. This is, perhaps, the most difficult part of universal love – keeping the offer of love open while defending yourself (or others) from someone trying to do you harm. There is, however, no alternative way to propagate universal love and the benefits this brings.

Honesty is a set of practices that are necessary for the development of trust both personally between individuals and abstractly between peoples. Without trust between people the full benefits of love cannot be realized nor can animosities be avoided. There are many actions defining honesty but the most universally required actions are:

- Adhering to the truth, both stated and implied.

- Not taking unfair advantage of others ("unfair" defined as not conforming to the established rules and/or expectations of the situation).

- Doing what you say you will do.

- Seeking possessions or rewards only in return for rightfully earning them (do not cheat, steal or use force against others to obtain desired possessions/rewards).

In the political realm honesty is always in short supply, which is why political compromises seldom succeed in the long term.

Humility, the state of being humble, is essentially acknowledging that regardless of your accomplishments or acquired status in society you are no better than others in the overall scheme of the universe. Typical dictionary definitions of humility are: "The feeling or attitude that you have no special importance that makes you better than others" or "simply the quality or state of not thinking you are better than other people".

Possessing humility does not mean denying that one may, in fact, be more intelligent, stronger, more attractive, more successful, etc. than many or even most other people and that you may have worked hard to develop these qualities and successes. The humble person simply recognizes that genes, the circumstances into which they were born and life's chances – all factors beyond the individual's control – played a large role in their life. Further, the humble person recognizes that traits, abilities and successes are but transient conditions that degrade with time and ultimately end in death, the fate of all life. Finally, the humble person recognizes that his/her life is but one brief contribution to the web of all life, the purpose of which only the God Power knows for certain.

Only by denying the logical consequences of True Reality can one not be humble. That many manage to succeed in such denial is a source of much strife in the world. Humility is an essential ingredient for long-term harmony between individuals.

Responsibility, as used herein, has a dual meaning: it is the personal commitment both to accept the consequences of one's actions and to do the right thing in all situations. In many respects "Responsibility" may be considered a subset of "Honesty" but lack of personal responsibility is so prevalent today that it warrants a separate emphasis.

There are many examples of personal responsibility in action with the following list comprising perhaps the most universally critical ones:

- You educate yourself as to what constitutes True Reality in order to base your perceived reality on truth. In today's world this will require learning beyond what is taught in most schools. Fortunately, the information necessary to construct True Reality is available in the public domain for those seeking it.

- You accept the consequences of your actions whether taken on your own initiative or as part of a group. This assumes that no creditable direct coercion caused the action. What constitutes "creditable coercion" is highly situation dependent. All individuals have free will in making decisions although such can generally be compromised with sufficient coercion. However, simply blaming your actions on upbringing, a bad experience or society is not a creditable defense absent a clear cause-and-effect link.

- You prepare yourself to earn your way in the world. At a minimum this involves taking the initiative in developing your knowledge and skills with which to acquire the necessities of life for yourself and whatever dependents you may have. You may not in fact be – or think you are not being – treated fairly (whatever the term "fairly" means) but this is a contra productive excuse for not acquiring marketable skills for earning a living.

-You make every effort to fulfill your responsibilities, whether such responsibilities were assumed on your own initiative or by accepting a responsibility advanced by others, (again this assumes that no creditable direct coercion caused the acceptance of that responsibility.)

At this point I conclude my review of the Elements of True Reality (Part 1 of this book). To reiterate: the objective of Part 1 is to present the core of True Reality in a manner that meets the criteria of being unambiguously based, clearly stated, logically consistent and independently verifiable.

You are the judge of whether this objective was met to your satisfaction. In any event, the Elements presented herein I offer as meeting these criteria while providing a perception of reality that is consistent with the universe that we experience. Should you believe otherwise then these elements should provide a clear target for your opposing arguments.

Part 2: Using True Reality to Understand and Address World and Personal Issues

Purpose of Part 2

(Applicable dictionary definition of "issues" : a vital or unsettled matter, concern, problem)

The purpose of Part 2 is to demonstrate how life's issues can be logically addressed within the boundaries of True Reality dictates. Actually, in most cases the application of True Reality dictates to issues will simply appear as applying common sense. That is how truth usually works.

True Reality provides the indispensable guide for both understanding and addressing issues. While we will never know with scientific certainty all the components of True Reality what we do know constitutes the only true reference for guiding the human mind.

This does not mean that using True Reality as the guide makes understanding issues immediately obvious or that a correct response is easily recognized, let alone carried out. It does mean that issues are a consequence of conflicts with True Reality dictates and that the correct resolution must be in compliance with these dictates. If the goal is a better world for individuals and societies then True Reality constitutes the boundaries within which both understanding and resolution must be found.

In what follows I offer examples of addressing issues from the perspective of True Reality. In keeping with this book's self-imposed constraint for brevity I make no attempt to address all True Reality conflicts that may be involved in an issue.

Rather, I address only the more significant conflicts with more or less obvious corrective actions. The actions offered herein as corrections may not be all that is needed to bring an issue into full compliance with True Reality but they are a necessary part of such actions.

From the many issues that exist I have selected a few of general concern which have corrective actions available that are not likely to generate such emotion as to disrupt objectivity. I recognize that establishing such a balance is questionable but that is the intent.

Most of the issues selected directly relate to the United States system of government, but the principles involved are applicable everywhere. Political issues were selected as political turmoil is much on people's minds and rational responses are in short supply. Political issues particularly need the guidance of True Reality in order to generate rational responses.

Examples

Purpose of the Individual in the scheme of "Everything":

A God Power created the universe and all life within it. Evolution appears to be the God Power's method for developing life. As the God Power set the rules of evolution at Creation all life must serve some purpose in that evolution, whether we currently understand this purpose or not.

In the human individual evolution has produced a life form far more advanced than any other. Mankind's physical form, intelligence and manual dexterity all serve to place Man in the dominant position within the hierarchy of all life forms. Further, Man is the only species intellectually able to contemplate the purpose of life; life after death (yes/no/what); and the existence of a God Power. The existential question for Man is:" What is the purpose of Mankind in the scheme of "Everything"?

An individual never knows for certain the true purpose of his/her life but nevertheless inherently desires the freedom to pursue what he/she thinks of as their purpose. In addition, all individuals possess the free will to act as they think best. The only God Power restraints placed on this free will are what True Reality imposes -- which restraints are considerable.

The individual acts in accordance with the dictates of his/her modified nature and perceived reality. However, only acts compatible with the dictates of True Reality have positive consequences in both the long and short term. False realities serve only to corrupt the individual's free will into taking actions that are ultimately detrimental, however attractive they may appear in the short term.

It is the individual's responsibility to seek out and embrace True Reality and it is Mankind's salvation that the individual does so. Such an undertaking is recognizably never-ending but the individual has no higher purpose in the scheme of "Everything".

Educating the individual so as to instill True Reality as this/her perceived reality is the only solution to this issue.

Inalienable Rights: Individual vs Group

True Reality dictates that individuals have inalienable rights that are universal for all of mankind. Groups have no inherent rights beyond what the individual members themselves possess. Groups are simply social constructions to better attain some common objective sought by the individual members.

The foundation of an individual's inherent rights can be expressed no better than the previously quoted phrase extracted from the preamble to the United States Constitution: "*We hold these truths to be self-evident, that all men are created equal, that they are endowed by their Creator with certain inalienable rights, that among these are Life, Liberty and the pursuit of Happiness*".

Although the above statement of inherent rights is vague on specifics, the beauty of it is that whatever laws are enacted to secure these rights, even if initially ineffective or even detrimental, eventually all will evolve into beneficial laws -- providing they are applied equally to everyone and end up being compatible with True Reality.

No law will long exist (in historical terms) if it is detrimental to those in power and/or conflicts with true reality. Failure to meet these two criteria -- equal application and True Reality compatibility -- is a sure signature of a law that is detrimental to an individual's inherent rights and, thus, ultimately detrimental to Mankind as a whole.

It is often declared that the "the rights of the many (think group) take precedence over the rights of the few (think individual)". This is a false vision of reality that allows groups (large or small), simply through association, to acquire "rights" that take precedence over the inalienable rights of other individuals. Historically it is this vision that "justifies" the "many" to dominate the "few" (If not outright dispose of them). Inversely, this vision also allows a "few" (dictators, cults, governments, etc.) to dominate the "many". Most of Man's history revolves around the problems caused by selectively deciding who has what rights.

The only solution to this issue is to enforce equal inalienable rights for all individuals.

Religion

From the individual's perspective, religion is essentially the sum total of a person's beliefs concerning:

- How the universe and life was created (God Power or otherwise)
- Whether there is an existence after death
- Whether the God Power remains active in our lives
- The nature of one's relationship with the God Power
- How one should personally behave
- How one should treat others.

These beliefs are established through acquired knowledge, life experiences and reasoning processes.

For most people their religion is first acquired through the teachings of a religious organization that they are exposed to as they grow. However, everyone, when their reasoning powers mature, should analyze any set of religious beliefs acquired in this manner to see how well they reflect True Reality. Most religions present today were started long ago when Man's knowledge and the demands for survival were much different from that of today. Such differences likely have little impact on a religion's fundamental belief in a God Power but one may find some beliefs, doctrines and rituals that are incompatible with today's understanding of True Reality. As True Reality is a manifestation of the God Power's plan for our universe, all religious beliefs, doctrines and rituals must be compatible with the Elements of True Reality

Unfortunately, religions, like any other organization of Man, are subject to all the strengths and weaknesses of the individuals involved, regardless of their stated purpose. Historically, one finds that some religions were created for the sole purpose of controlling large populations for the benefit of a relative few. Even where the tenants and dictates of a religion are universally beneficial for mankind, some individuals -- alone or in association with others -- will find ways to corrupt it for their own benefit should the opportunity present itself. A firm knowledge of True Reality is the only effective defense against religious manipulation.

A quick test for any religion is how well it universally promotes harmony between individuals. For a religious organization -- whose reason for being is based on belief in a God Power that created all of Mankind -- perhaps the foremost True Reality dictate is that all of mankind is equal before the God Power, *endowed by their Creator with certain inalienable rights, that among these are Life, Liberty and the pursuit of happiness.* Reason alone is sufficient to confirm that universal adherence to the above "inalienable rights" is a prerequisite for maximum harmony in the world. Anything less inherently produces conflicts between those possessing and those denied such rights.

Education

The individual is born into a world where competing interests with differing agendas immediately seek to educate him/her in a manner that promotes one or another of these agendas. The objective is to shape the individual's modified nature and perceived reality to produce behaviors and actions that support a particular agenda. Through control of the education process such shaping is easy and effective, even if the perceived reality being promoted is false. What is not taught is as important as what is taught and most education programs today do not teach True Reality. This is intentional as no false reality can survive in direct competition with True Reality when the "Truths" on which each is based are known. Teaching individuals the basics of True Reality should be a central goal of education.

To achieve maximum harmony in the world it is necessary for True Reality to be the perceived reality of individuals. There is no logic in assuming otherwise. One is unlikely to make correct decisions based on false premises. As the saying goes in relation to computers, "garbage in garbage out", so it is with individuals -- unless they have a point of reference that allows them to recognize garbage. That essential reference is what True Reality provides.

There is no defensible position that proposes that the individual should be educated in anything other than True Reality. Logical differences can only arise over what constitutes True Reality. The Elements of True Reality I have extracted from public knowledge and presented herein are offered as the individual truths that form the core of True Reality. Thus, these elements should be taught, complete with all supporting information, to all individuals as part of the education process. Those who disagree -- meaning those promoting other "realities" - must be required to identify the elements ("truths") supporting their perception of reality in order to compare them with the elements presented herein. Our perception of True Reality can only benefit from such comparisons.

In practice the proponents of false realities seek to discredit the Elements of True Reality but not by direct confrontation with their own elements of reality. Instead, they construct a cloud of false, misleading and biased information on which to rest their "Reality". Such information, presented continuously through both formal education and public communications, is effective in instilling false realities in the unprepared individual. Thus, the individual's best defense of such tactics is to learn how to recognize false, misleading and biased information.

Critical reading and listening skills provide the most effective general tools for identifying information that is misleading or biased Recognizing outright lies still requires knowledge of the truth but for recognizing misleading and biased information -- the bulk of what is communicated today -- critical reading/listening skills are often sufficient. Such skills should be taught to the individual as part of his/her formal education but such is nowhere to be found in public schools and seldom in private schools. The reason why these skills are not taught is precisely because they are effective in arming the individual against misleading and biased information. Those interested in spinning false realities cannot risk having such tools in the hands of the "average" individual.

Critical reading/listening is a subset of critical thinking, a skill that involves life-long learning, best started at an early age. However, there are some simple techniques that can be eye opening to the individual in demonstrating just how pervasive false, misleading and biased information is in public media.

One such technique is to take any written article such as a political article in the newspaper (admittedly a worst case example) and cross out all adjectives and adverbs involved to see what information remains. What remains is the bare information contained in the article, shorn of the author's interpretations and biases, frequently leaving not much of value. If what is left could have value - should it be true - a second technique comes into play.

The follow-up technique is to ask yourself whether the author can be reasonably expected to know what he/she is relating, either through direct experience/observation or through supporting "facts". If this reasonable expectation cannot be justified then the entire article has an objective that does not include conveying truth.

If the article fails both tests then the only thing of value that a reader can extract from the article is to recognize that the author does not let truth stand in the way of promoting an agenda. This recognition forewarns one when encountering additional information from this same source.

Because of the impact education (both formal and informal) has on an individual's modified nature and perceived reality - and thus his/her actions - educating in accordance with True Reality is a necessary step on the path leading to a better world.

The United States Constitution

In the history of Mankind the United States Constitution is, in its original intent, the purest governmental attempt to empower the individual in securing his/her inalienable rights as established by natures laws. The Founders recognized that governments represent the greatest threat to individual freedoms and sought to give the individual the means to control this threat. This approach was also in compliance with True Reality -- whether they thought of it in this way or not -- in that they saw the individual as the key to world harmony.

Unfortunately, the safeguards put into the Constitution to control government powers, while balanced in principle, proved inadequate in practice. Many of the Founders recognized that the safeguards had weaknesses but such was the best they could accomplish given the political situation at the time. To quote John Adams: "Our Constitution was made only for a moral and religious people. It is wholly inadequate to the government of any other." This weakness was all that was needed for "others" to begin subverting the constitution for their own purposes.

Our country's moral and religious underpinnings have slipped since the time of our government's founding. Various special interests have been chipping away at the Constitutions original intent from the start, and these efforts have accelerated in recent times. There can, of course, be valid reasons for changes to the constitution. The Founders anticipated the probable need for such change and established a procedure to do so. This procedure preserved the individual's power to initiate and/or approve such changes.

Presently, the individual's power to initiate and control changes to the constitution has been largely nullified by governmental end-runs around constitutional procedures. Changes instituted by circumventing constitutional procedures invariably increase the government's power over the individual. Evading constitutional procedures is an open admission that the changes sought could not be accomplished in open debate. Even should the short-term benefits of such evasion appear desirable to some, the long-term consequences of such power over individual rights are never good (as history demonstrates).

To bring the government back to the original intent requires plugging loopholes in the constitution so that people who are not "moral and religious" cannot so easily corrupt it. This plugging of loopholes will require instituting governmental operating rules and procedures more in tune with True Reality. Some of the more obvious changes needed are discussed in the following examples.

Controlling the "People's" Representatives

In the early years of our constitutional government the people's representatives -- Presidents, Senators, Representatives, Judges, etc. -- were most often individuals whose character and accomplishments were well known to the electorate and who saw service in the government as a duty. They did not seek nor expect such service to be a career in itself nor to be financially rewarding. Power over others and the potential for financial benefits was a likely attraction for some but the opportunity for such abuse was limited by the then close adherence to the constitution and by the vigilance of the people when voting. Much has changed since then.

Today the "people's representatives" have transformed our governmental system into something in which service can be financially attractive, particularly as a career, and they have made a career path more feasible by, among other things, manipulating the election process so as to enhance their continual reelection (more on this to follow). The election process itself has become so costly that the already rich candidates enjoy a decided advantage over less wealthy opponents. Add in the growing disregard for the constitution's original intent and the door is open for the powerful to support candidates that will promote their various agendas. The Constitutional checks and balances no longer reliably control our governmental representatives.

It is no surprise that the original intent of the constitution has degenerated in this way. True Reality guarantees that such would eventually happen. If a situation exists that can be exploited to someone's advantage, that situation will ultimately be so exploited (Psychological Derivative, Element 2). The constitutions reliance on a "moral and religious people" to maintain control over the government provided just such a situation.

It is no accident that as the role of morality and religion in our society declines, the corruption of the constitution increases in lockstep. Rather, this is the result of a carefully planned and executed long-term effort to promote a false reality in which morality and religion don't matter or may even be detrimental to the well-being of mankind.

To create a perceived reality in which everything is relative and the government is the final arbiter for what is best for the individual requires that morality and religion become irrelevant in people's thinking. Efforts to marginalize morality and religion through education and social media have been underway for a long time. These efforts are now bearing fruit in people who do not perceive a True Reality. Such is the age-old path leading to societal dissolution and the loss of individual freedoms, enforced – as usual – by a government.

For the individual (in aggregate) to regain control of government he/she must begin by embracing True Reality. In no other way can the individual obtain the knowledge and perspective for deciding both what the proper relationship between the Individual and the Government should be and what the requirements must be for establishing and maintaining this relationship. The Founders of our Constitution did a remarkable job of defining a True Reality relationship between the individual and the government (see the Preamble to the Constitution) but failed to establish adequate controls over the actions of our representatives in government. However, there is nothing except the will to do so that prevents "We the People" from adding the missing controls.

Of course, those who benefit from the existing government structure will resist any controls that threaten their power and the benefits that this power provides. They have been successful in corrupting the original intent of the constitution by patiently constructing false realities that appear attractive to many people. Individuals that have True Reality as their perceived reality have no choice but to be equally persistent in promoting the controls necessary to bring the government back into compliance with our constitutions original intent.

The problems with our current government are neither complex nor difficult to understand. To the contrary, the shortcomings and corruptions are obvious and constantly on display. There are basically four areas of control that are needed to bring our government back to the vision encapsulated in our constitution: term limits, logical operating rules/procedures, transparency and appropriate compensation for service.

Term Limits

It is to be expected that elected governmental positions that confer power, and/or are potentially profitable, will attract some individuals that wish to use those positions for their own benefit. It is the people's (electorate's) responsibility to prevent such individuals from attaining government service. Unfortunately, experience has shown that individuals who do not embrace True Reality (and there are many) are easily misled into electing just such individuals.

It is also to be expected the even when individuals of demonstrated character are elected such individuals often change when actively in service. Over time, the subtle effects of wielding power over others, the collecting and allocating of large sums of money not your own (even when doing good) and being constantly stroked by various supplicants seeking favors takes its toll. Few individuals can withstand long-term service without eventually believing he/she does indeed deserve some special benefits, does deserve to be reelected and does indeed know what is best for the people. Whoever coined the maxim "The road to hell is paved with good intentions." may have had such politicians in mind.

Term limits provide a practical control for limiting the damage from both the premeditated self-serving elected official and the corrosion of a person's character that long-term service engenders. Neither will correct themselves while both will point out that the electorate now has the power to limit terms through the ballot box. Therefore, they contend that term limits are unnecessary and actually prevent the electorate from keeping a representative they like.

Of course, politicians know that the electorate can be manipulated by biasing the media, by keeping them (the people) ignorant of what they (the officials) are actually doing, by pitting one group against another, by rewarding their electorate with someone else's money and by creative mapping of voting districts — to list a few of the methods currently employed to keep themselves in office.

The average individual has neither the time nor inclination to independently verify the true merits of election candidates and, thus, they must largely depend on public media to provide such information. Manipulate the media — the career politician's strong suit — and you manipulate the average voter. Term limits make such manipulation mute at some point in the election cycle. By so doing the attractiveness of political positions to self-serving individuals is greatly reduced. An added benefit is that the power base of "establishment" politicians is also greatly diminished if not eliminated.

No government Official is irreplaceable. Some individuals' standout because of their integrity and effectiveness in getting things done but there are always others who can take their place if the voters are vigilant. In any event, such "standout" individuals can readily find other government positions in which to serve if that is their inclination. Term limits are the people's single most effective measure for regaining control over our representatives.

Congressional Rules and Procedures

By itself term limits only go part of the way toward disrupting the fiefdom that most of our elected officials actively promote. Changes also need to be made to the rules and procedures that each branch of government has established over the years — without any input from the electorate. The current rules and procedures best serve their own interests, not necessarily what is best for the people. In addition, apparently everything can be bent to the interest of the majority at any time. They play power games amongst themselves with the electorate simply along for the ride.

The www.senate.gov website for "Rules and Procedures" begins their documentation of such with this paragraph: *"The legislative process on the Senate floor is governed by a set of Standing Rules, a body of precedents created by rulings of presiding officers or by votes of the Senate, a variety of established and customary practices, and ad hoc arrangements the Senate makes to meet specific parliamentary and political circumstances. Knowledge of the Senate's formal rules is not sufficient to understand Senate procedure, and Senate practices cannot be understood without knowing the rules to which the practices relate."*

In perusing the specifics following this statement I found the above paragraph to be an apt description of the existing state of affairs. In other words the Senate rules and procedures provide something like guidelines that are not particularly binding unless the majority decides that they are in specific situations.

Such operating flexibility is a major source of the power games legislators' play. Applying logic in the interest of effectiveness appears to be largely missing in the existing legislative process.

We can do better. A set of operating rules and procedures that are logical, fair and <u>binding</u> can be formulated to obtain what is best for the people rather than what is best for legislators. The political establishment has much invested in making the present system work to their advantage and will fight change every step of the way, but such change is clearly necessary.

Congressional Operational Transparency

Except in matters of national, state and/or local security, full transparency of governmental operations is key to informed voter participation in how we are governed. Even amid security concerns the maximum information that does not aid the enemy (criminal, suspect, etc.) should be made available to the public in a timely manner. In the past such transparency was technologically and logistically very difficult if not impossible to achieve, even if those in power wanted to do so. In the present digital age much greater transparency is feasible but still not likely to happen without a battle. Those in power seldom like transparency in their operations.

The people will likely have to force greater transparency onto government operations. However, some increase in transparency can be accomplished through websites maintained by the private sector. Such websites would initially involve taking existing government public records, often recorded in little known locations, and publishing such information on a readily accessible and trusted website. Because government information is often shrouded in legalese language some parsing may be necessary to convey the essence of what it means- thus the need for establishing trust in the website.

By itself such a website could only dig up the currently available government information and convey it to the public in a timely and comprehensive manner. At the minimum this information would enable the public to respond more quickly to government actions, perhaps before the action is completed rather than afterwards, as is often the case today. As the public understands more of what the government is doing there will be increased demands for more governmental transparency.

The end result will be a shift toward a government more in tune with people's best interests. Of course, the funding for any such public service websites would have to come from non-government sources and a public record maintained of these sources to ensure compliance.

Compensation for Elected Officials

To minimize corruption the founders of the constitution recognized that the compensation of elected officials should not be so financially rewarding as to provide a reason in itself to seek service. Ideally, compensation in all forms should create neither a disincentive nor incentive for serving. What exists today diverges significantly from this ideal.

Current base salaries of approximately $174,000 for House and Senate members, $249,000 for Supreme Court Justices, $231,000 for Vice President and $400,000 for President are not insignificant. These salaries place them well within the top five percent of all wage earners (the top one percent for the President). However, for an honest House or Senate member who doesn't want to make a career out of his/her service, and who has a family to maintain in his/her home state, this salary may well serve as a disincentive to serving.

A wealthy member would find such compensation immaterial to his/her needs while someone seeking a career in politics would find it an acceptable price to pay for future rewards. The result is that wealthy and career politicians (often one and the same) comprise the most numerous group within the House and Senate membership.

This bias toward rich and/or career politicians in government acts to mute the voices of the people, the majority of which are neither rich nor desiring a career in public service. Thus, the actions of government often do not reflect what the people as a whole desire. The lessons of history (which our Founding Fathers well knew) demonstrate that rule by an elite always degenerates into something that is bad for people outside the elite. The solution is to obtain a more representative distribution of views in our elected officials. Compensation plans that better reflect the actual cost of service to the elected individual will encourage more diverse and competent individuals to seek office.

The combination of term limits, more effective operating rules and procedures, greater operational transparency and adequate compensation for our elected officials would result in a government much closer to the ideals of our constitution.

Controlling the Supreme Court

The Judicial Branch of our government was established to serve one purpose and that purpose was to ensure that the laws formulated by the Legislative branch and the actions of the Executive branch do not violate the intent of the constitution. The power to do this was invested in nine Justices who are appointed for life terms. To fill vacancies the President submits candidates but the Senate must confirm them for this position. The responsibility to select Supreme Court Justices was split in this manner because of the exceptional power placed in the hands of just nine people.

The rationale behind giving the justices life appointments was to shield them from political pressures. The assumption was that the justices would be selected on the basis of demonstrated moral and religious integrity and whose oath of office to interpret the constitution as written and intended could be relied upon.

In the beginning there was a ready supply of such men (no women allowed at the time) as many had lived through the birth of our nation and well knew the intent of the wording, including that of the early amendments. As time has passed it has become more problematic to find and appoint such men – and now women – with similar integrity and dedication to the intent of the constitution.

Today there is not even a pretense that the selection of judges is anything but political. The politics revolve around whether the constitution should be interpreted as per the original intent or whether judges can interpret according to their perception of what is relevant today. This difference in political interpretations is of great importance.

Once a judge or judges are allowed to stray from the intent of written law then they bestow upon themselves the power to create laws without input from Congress, the states or the people. Placing such power into the hands of just nine individuals -- insulated from accountability by life appointments -- is placing too much faith in both the knowledge and innate goodness of Man.

Unfortunately, such politics are not confined to the Supreme Court; the appointment of all justices now involves this political contest. Without the moral and religious integrity so relied on by the Founders judges have been unilaterally amending the constitution and doing so with impunity.

To prevent constitutional amendment through judicial dictate Supreme Court Justices must, in some manner, be held accountable for judgments in conflict with the original intent. Should they believe current conditions require a judgment in conflict with the original intent they are now free, via existing constitutional procedures, to petition Congress and the states for an appropriate amendment to the constitution.

Should Justices make a judgment in conflict with the original intent, refusing to seek an enabling amendment, then an avenue must exist to nullify the judgment and remove the judge(s) for malfeasance. To date, efforts to nullify Supreme Court decisions by states and others have all failed, rejected as illegal by – you might guess – Supreme Court Justices.

The power to nullify a Supreme Court decision or to remove a Justice for legal malfeasance may well require a constitutional amendment to accomplish.

The most likely amendment would be to give a coalition of Congress, the states and the people this power. In any event, renegade Justices must be controlled to protect both our "inalienable rights" and the Constitution's Bill of Rights that they are unilaterally abridging. In fact, our entire system of freedoms and laws are in jeopardy if "judicial relativism" remains unchecked.

Concluding Statement

In order to not further distract from the primary focus of this book, which is to present the core structure of True Reality, I forego introducing more examples of analyzing issues from the perspective of True Reality.

I conclude by offering the following paragraph as a concise summary of the role that True Reality plays in Man's affairs:

To be successful the solutions to all issues facing us today and in the future must conform to the dictates of True Reality. Any nonconforming "solutions" are inherently illogical and must fail. They are illogical because the consequences posed by True Reality cannot be circumvented. True Reality is simply the existence that the God Power has created for us and for everything else. The best we can do is to seek an understanding of this Reality so that we can comply with its dictates.

Addendum

List of Elements of True Reality

Fundamental Elements of Physical Reality

1. The origin of the universe is forever <u>physically</u> unexplainable.

2. Everything in the universe consists of energy in one form or another.

3. The fundamental nature of energy is unknown; only <u>some</u> of the forms that energy can take are known.

4. The original source of all energy is unknown.

5. What we perceive as solid matter is almost entirely empty (of matter) space.

6. Empty (of matter) space is filled with energy.

7. The passage of time is not everywhere the same.

8. All life is initiated by, and dependent upon, the consumption of energy.

9. The evolution of all life forms is an established fact but the force(s) driving evolution remain unresolved.

First Derivative Elements of Physical Reality

1. All of Man's understanding of the universe is limited to the "how" not the "why" of its operation.

Fundamental Elements of Psychological Reality

1. All individuals are born with a common set of characteristics/instincts.

2. At birth no inherited characteristic/instinct is expressed equally in all people. The intensity of <u>expression</u> for any individual characteristic/instinct can range from extremely weak to obsessively strong.

3. The <u>expression</u> of an individual's inherited characteristics/instincts (nature) is subjected to modification immediately following birth (perhaps before birth) through the interaction of four inputs: acquired beliefs, experience, knowledge and reasoning.

4. An individual's <u>perceived reality</u> is constructed from the interaction of five inputs: nature, beliefs, experience, knowledge and reasoning.

5. Except in cases of physical brain defects the individual always behaves/acts in accordance with the dictates of his/her nature and perceived reality.

First Derivative Elements of Psychological Reality

1. In any sufficiently large population of people an extreme expression of all inheritable characteristics/instincts will be found to exist.

2. If a situation exists that can be exploited to someone's advantage, that situation will ultimately be so exploited.

3. All organizations/associations -- organized religions included -- are in operation social constructs subject to all the strengths and weaknesses of the individuals involved.

4. Shaping the perceived reality is the central battle between competing interests.

5. Some competing perceived realities cannot forever peacefully coexist.

6. Violence always has and always will play a role in man's interactions, employed by some for advantage, required by others for defense.

Transcending Elements of Reality

1. There is a God Power that created "Everything" and now guides the evolution of that "Everything", which includes the universe and all life within it.

2. All God Power relationships with humanity are with the individual.

3. The individual is the template that mankind follows.

4. Only perceived realities founded on true reality can create the conditions for a better world.

5. Just four attributes of an individual's modified nature are needed to achieve maximum harmony in the world: Love, Honesty, Humility and Responsibility,